TOXIC WAIST?…
Get To KNOW SWEAT*!*

Built by Denny Miller
Illustrated by Mike Royer

TOXIC WAIST?... Get To KNOW SWEAT*!*
By Denny Miller
Illustrated by Mike Royer

Printing History
First Edition, April 2006

Manufactured in the United States of America
ISBN 0-9753917-1-2
PCN Locator Number 2005908746

Published by

TO HEALTH WITH YOU PUBLISHERS
8550 W. Charleston Blvd. #102-374
Las Vegas, NV 89117

Layout Design
Mike Bifulco

Front Cover Colorization
Tom Luth

Back Cover Design
Rob Ryder

Here's what some people have already said about TOXIC WAIST*!*

"Denny Miller definitely captures our attention with his long held beliefs about improving health and fitness through the use of reasonable exercise and diet commitments. This compendium of no nonsense, informational bites draws dramatic attention to a serious threat to the future of our society. Each of the points that is made, whether it is funny, sad, irritating and/or stimulating is worthy as food for thought about our thoughts on food and exercise. The use of light as well as heavy humor to make important points has long been part of the fun associated with reading his words. The illustrations are on target and provide good reinforcement to the messages followed in the text. The simple message 'TOXIC WAIST?…Get To KNOW SWEAT*!*' if followed, would make a positive difference to a population that is slowly and surely eating its way into the ground."

Dr. Glen H. Egstrom
Professor Emeritus
University of California, Los Angeles
Department of Physiological Science

"Inactivity, overweight, and obesity are growing concerns in all developed nations. The problem needs to be acted upon at all levels – individual, corporate and government.

Miller has taken a light-hearted approach to this weighty issue by assembling a fun-tastic collection of quips, quotes and cartoons. Not only will the immediate laughter be good for your health, but you can use the ideas to help prevent toxic waist at home, in school and in your community.

Start to 'know sweat' by skipping back and forth between your favorite sections; then jump to your own conclusions. If these exercises don't work out for you, take the book on a daily two-mile walk with a friend."

Thomas L. McKenzie, Ph.D.
Professor Emeritus
Dept. of Exercise and Nutritional Science
San Diego State University

"Miller's book may prove to be the motivation that you need to begin an exercise program. After reading 'TOXIC WAIST' you will either sit down and grow in all the wrong places or get off your duff and exercise. This book is filled with good information, catchy sayings and humorous illustrations. Reading this book may be the push you need to take the necessary steps to enjoy a better and healthier lifestyle."

Roger Dickinson
Executive Director
Indiana Basketball Hall of Fame

"Denny Miller has provided a very entertaining and informative book on a healthy lifestyle. More important than the book, we must put the principles contained in it into action in our every day lives.

The principles contained in 'TOXIC WAIST?…Get To KNOW SWEAT!' are extremely important to all Americans! Miller's unique and clever writing style make the book an enjoyable read as well as a great motivational tool to get our bodies active.

I'm particularly pleased to see the dedication of this important work to Denny Miller's father, Dr. Ben W. Miller, who was a distinguished graduate of Indiana University and served on the faculty for a time. He went on to an important international career, and it is obvious that he had an impact on his son. Dr. Miller would have been very pleased that his son has carried on his important work."

> Tony A. Mobley, Ph.D.
> Professor Emeritus, Indiana University
> Executive Director
> National Recreation Foundation

"In 'TOXIC WAIST?…Get To KNOW SWEAT!,' Denny Miller hit the nail on the head. With a healthy blend of satire, the positive, and cold facts, he leads the reader to a commitment to life-long exercise, the only lasting, and true road to fitness and health."

> Swen Nater, Author, Poet, Motivational Speaker
> Assistant Sporting Goods Buyer, Costco Wholesale
> NBA First-round Draft Pick, 1973
> ABA "Rookie of the Year," 1974

"Toxic Waist?…Get To KNOW SWEAT! is really what it is all about. In a country where more than 50% of its population is overweight, it is a reminder to us all that it is time to take responsibility for ourselves. Denny Miller has gathered many great quotes and cartoons to illustrate where we are, but more importantly, where we need to be. He has made it simple, direct and to the point. We must find movement that we enjoy and then get moving."

> Dan Latham
> Cyberobics, Inc.
> Fitness Consultant

"Being from Brazil, I look at the citizens of the United States with a point of view that is not blinded by local myths. Example: a U.S. teenager told me, 'Exercise would be bad for his heart.' He really believed that exercise could be very harmful. And, why not? He has grown up in a school system, for the most part, that did not include Physical Education in the curriculum.

The result of not teaching the importance of the mind-body connection, creates a population with no idea that their body's level of fitness is their responsibility. This lack of knowledge and misinformation are key reasons for this obesity epidemic. This book is not the solution to the Obesity Epidemic but it is a good start."

Paulo Figueiredo
Masters Degree in Physical Education,
Long Beach State University
Former Brazilian National Champion, Swimmer
Former Chancellor Officer – Diplomat for Brazil

"Denny Miller has written a fun book that sends a powerful message. The many quotes and cartoon illustrations will get the attention of the reader. Denny's book will challenge people to evaluate their lifestyle in terms of diet, health and the value of exercise. I highly recommend the book to those seeking a healthier life.

Dr. Gary Cunningham
Athletics Director
University of California, Santa Barbara
Former Head Basketball Coach
University of California, Los Angeles

"Toxic Waist?…Get To KNOW SWEAT! takes a fun, humorous approach to a very serious public health threat. The use of short, simple messages with cleaver drawings makes the book very readable. Hopefully, this book will help motivate some of the population to eat healthy foods and get more active!"

Heather Robinson, MPH
Lifestyle Coach for Health and Fitness

This book is dedicated to my father,
Ben W. Miller, Ph.D.
Professor Emeritus, UCLA

A sound mind in a sound body. Or in today's language, "An affirm mind in a firm body." Both describe the Golden Age of Greece. That was my dad's goal. He wanted to see that Golden Age happen in the United States. He thought the way to accomplish that was through education – in school and with the parents.

To reach his goal, he worked very hard to achieve positions of authority where he could influence teachers, administrators, politicians – including Presidents and community leaders; people of power who could change policy and educational curricula.

He was Chairman of the Department of Physical Education at UCLA for more than ten years. He served on Presidents Eisenhower and Kennedy's Advisory Committee on Youth Fitness. He was President of the American Academy of Physical Education. He was a charter member and Trustee of the American College of Sports Medicine and a Fellow of the American School Health Association and of the Public Health Association. He was President of the American Association for Health, Physical Education, and Recreation.

His career in national and international organizations allowed him many opportunities to travel. He gave over two hundred-fifty addresses in the United States and abroad, fifty of which were published. He attended all of the Olympic Games from 1932 to 1984 and visited more than eighty countries, gathering information about the history of physical education and sport. He created a class on this subject, which he taught at UCLA. He also organized and was a member of the North American Society of Sport Historians.

In 1943 he wrote "PHYSICAL FITNESS FOR BOYS," and dedicated it to "The physical fitness leaders of American youth."

Fitness is no big mystery. It's not hard to understand how to be physically and mentally fit. My dad did his best to yell that message from the roof tops. I am sure Dad would want me to pass the word along. Dad's goal of an affirm mind in a firm body was a great idea. It still is *!*

~ D.M. ~

President John F. Kennedy with Dr. Ben W. Miller (Dad) in the White House

ACKNOWLEDGMENTS

To do the research for this book, I used over one hundred books, periodicals, studies, surveys and interviews regarding the current level of physical fitness of individuals in the United States, in age groups including preschoolers to people in their nineties.

The information is mostly from scientific studies done by Medical Doctors, Nutritionists, Physical Educators, University Departments, State and Federal Officials, including five U.S. Presidents, the Surgeon General and State Governors, authors of books, papers and articles on health and fitness. Professional organizations such as AAHPERD (American Association of Health, Physical Education, Recreation and Dance), Fitness Instructors, professional, intramural, amateur, inter-scholastic and weekend athletes, and health-minded corporations. The list also includes producers of television shows that care about our children's health such as "Sesame Street" and "Lazy Town," and healthy individuals from various age groups. These individuals are listed below in the order their names appear in the book and are in the Bibliography along with their publications.

Many thanks to these dedicated and caring professionals! They have proven that a healthy lifestyle, in great part, requires that YOU get to KNOW SWEAT! They are:

Ben W. Miller, Ph.D., President John F. Kennedy, President Dwight D. Eisenhower, Dr. Laurence J. Peter, Jay Leno, Daniel Costello, Governor Mike Huckabee, President Bill Clinton, Steven R. Covey, Coach Frank Broyles, Dr. Isadore Rosenfeld, Helen Keller, President Franklin D. Roosevelt, James Brown, Robert Grudin, Surgeon General Vice Admiral Richard H. Carmona, M.D., Greg Critser, Steve Allen, Dr. Spencer Johnson, Garrison Keillor, H. G. Wells, Ralph Waldo Emerson, President Theodore Roosevelt, David Letterman, Dr. Trevor Orchard, O. Carl Simonton, Bill Cosby, Seneca, T. S. Elliot, Norman Cousins, Dr. Phil McGraw, Miriam Nelson, Ph.D., Bob Greene, Governor Arnold Schwarzenegger, Dr. Jules Hirsch, Ellen J. Langer, Dr. Martin Seligman, Dr. Andrew Weil, Orson Welles, Ambrose Bierce, Bob Anderson, Thomas Henry Huxley, Dr. Herbert Benson, Dr. Marilyn Gellis, Dr. Seuss, Jack Paar, Phil Silvers, Julie Andrews, William Shakespeare, Cicero, Thomas McKenzie, Ph.D., Dr. Wayne Dyer, Dr. Hans Selye, Coach John Wooden, Anthony Newley, Senator Dennis Nolan, Dr. Robert Ross, Maria Shriver, Senator Martha Escutia, Senator Abel Maldonado, Kim Belshe, Mihaly Csikszentmihalyi, Rosemarie Truglio, Magnus Scheving, Dr. Thomas N. Robinson, Don McGuire, Cyma Zarghami, Lynn Smith, Governor Jim Doyle, John Burke, Melissa Johnson, Dr. Leonard Epstein, Dr. Fay Boozman, Jim Hill., Lynn Swann, Dr. Julie Goldberg, Dan Latham and Montagne.

I also want to thank the students in my fitness classes for showing me, by their gains in strength, endurance, flexibility and balance that to "Know Sweat" is essential to improve one's health and quality of life!

Thanks to George McWhorter, Curator of Rare Books, Ekstrom Library, University of Louisville, who has edited both of my books. He also edits his own magazine, "The Edgar Rice Burroughs Bulletin." He has the strongest TARZAN YELL ever.

Henry Unger was responsible for the "Know Sweat!" wrist bands. Hopefully these sweat bands and the active/exercising people wearing them will Know Sweat!

Rob Ryder, Graphic Artist, designed the back cover; Tom Luth, Digital Artist, colorized the front cover; and Mike Bifulco, Graphic Designer, prepared the book for Sheridan Books, Inc. Their contributions are worth a thousand words. Thank you.

My wife Nancy has read and reread the book at least one hundred thirty-eight times and swept out all the typos. She contributed more and better ideas on the book's structure and content than I did. She also held classes of one to help me down off the wall and get back to the task at hand.

Michael Royer created forty-two full page illustrations and the front cover. His imagination added power and humor to my thoughts. His humorous drawings make it easier to "swallow" the sometimes bitter message of this book. If any author needs a talent to make his/her book better, do not hesitate to call on Mike.

I will always be in debt to Mike Hibler, my attorney and former UCLA Basketball teammate and road trip roommate. "Big Mike" said "Get Royer for your book. He's drawn for Disney for twenty-one years and he's the best." My "old roomy" was right!

And thanks to Dad, Dr. Ben W. Miller, Ph.D., Professor Emeritus, UCLA Physical Education Department. I grew up with the best mentor in the country, in my life time, in my home!

To get a better view of things
some people need to stand on the shoulders
of Giants that came before them.
I am one of those people.
Thank You to ALL the Giants listed here.

WARNING

The author is not rendering medical or professional services. If an individual is considering changes in his or her health-related practices or regimens, then medical advice or other expert assistance is required, and the services of a competent physician or other professional person should be sought.

PREFACE

How many books on fitness have been written? Hundreds? Thousands? They fill up shelf after shelf in every book store and library in the United States.

How many newspapers across the country have at least one article on fitness every week? Almost all of them.

How many magazines include articles on fitness, nutrition and diets every week? Most of them.

How many fitness videos are available? How many syndicated TV exercise shows can we turn on?

How many gyms, personal trainers, nutritionists, physical educators, playing fields, parks, hiking and biking trails, swimming pools, physicians, physical therapists, sporting goods stores and spas are there to help us???

PLEASE TELL ME
WHY ARE WE ONE OF THE
FATTEST PEOPLE
IN THE WORLD*!*

This book's purpose is to irritate you, shame you, make you angry enough, frightened enough, sad enough and care enough to motivate you to take your condition seriously. Get off your butt and on the move*!*

The human body was made to move. If you don't move you will be covered…first by shadows; then by sheets. And, *last and too soon by shovels.*

Being fat is more deadly than smoking!

Get moving. It's up to you.

For years fit, healthy people in our country have been the brunt of jokes. Ever hear… "You exercise, keep fit, die young and have a beautiful corpse."

The "Fit People" are the minority now. It's their turn to whine and make fun of people in the "Fat Majority." Just the other day I heard this remark:

>"*Stay fat,*
>*Die young and*
>*Leave more room for the rest of us.*"

It's not easy being in the majority.

"There comes a time in the affairs of men when you must take the bull by the tail and face the situation."

~ *DR. LAURENCE J. PETER*
"THE PETER PLAN"

HIND PLIGHT

Paramedics and health care facilities are using wider and stronger gurneys to accommodate fat folks.

Airlines require fat people to supply their own seat belt extensions because the existing seat belts are not long enough to buckle around huge toxic waists.

A company in Las Vegas that sells "Big John Toilet Seats" has been doing a *big business.* The "Big John" seat is *five inches wider* than the standard size.

The demand for extra large coffins is at an all-time high.

Jay Leno, TONIGHT SHOW Host, read a newspaper ad on one of his shows, for *"waistband extenders"* that connect to your waistband button for added girth.

KNOW SWEAT!

Excerpts from an interview done by Los Angeles Times Staff Writer, Daniel Costello...

"I was standing at the starting line of a marathon recently and a woman came up to say that I had inspired her to lose weight and to exercise. She had lost one hundred pounds."

Why did the woman thank Arkansas Governor Mike Huckabee like that? Because the governor had lost one hundred-ten pounds in the last two years. He has gone from a life-long couch potato to a forty-nine year old guy in better shape than he was at the age of eighteen. He went from a two hundred-eighty pound "slug" to a trim one hundred-seventy pound marathon runner.

Why did he inspire that woman and hundreds like her? Because he had written a book, "QUIT DIGGING YOUR GRAVE WITH A KNIFE AND FORK," that includes a twelve stop plan to help overweight people slim down. That's why*!*

He has become a very influential *Role Model!* "If I can get healthy anyone can." says the Governor.

He had impressed former President Bill Clinton so much that he asked Governor Huckabee to join with him, along with the American Heart Association, in a new, national campaign against childhood obesity.

Governor Huckabee will serve as the Chairman of the National Governor's Association. Their mission for 2005-2006 is to improve the nation's health. Two years ago, while he was beginning his own journey to wellness, he helped form a state-wide HEALTHY ARKANSAS Campaign.

Arkansas is the only state doing body mass screening on every student. The results of the screening are mailed home to the parents.

The state put in more walking trails and opened Patient Information Centers for people with chronic diseases to learn how to get healthier.

There are "Roll Models" – too many. And then there are "Role Models." Governor Huckabee is a *"ROLE MODEL"* for all of us fighting the battle of the bulge, which is the nation's obesity epidemic. As Chairman of the National Governors' Association, *I am betting he will convince other Governors to follow his lead.*

POWERFUL LESSONS IN PERSONAL CHOICE

"The way we 'see' (understand) the world is called a *paradigm*. We simply assume that we see things the way they really are. And our attitudes and behaviors grow out of those assumptions. The way we see things is the source of the way we think and the way we act."

~ STEVEN R. COVEY
"THE 7 HABITS OF HIGHLY EFFECTIVE PEOPLE"

Do you think being a *"roll model"* for your children is a powerful lesson?

"Governor, I found years ago that nothing ever tastes as good as it feels to be healthy." says Former University of Arkansas Coach Frank Broyles, who led the "Razorbacks" to a Championship Season.

~ ARKANSAS GOVERNOR MIKE HUCKABEE
"STOP DIGGING YOUR GRAVE WITH A KNIFE AND FORK"

8

HABITS LEARNED IN CHILDHOOD SET THE STAGE FOR GOOD OR POOR HEALTH IN ADULT LIFE

The September 25, 2005 , Las Vegas Review Journal PARADE Section ran "HEART HEALTH SHOULD START EARLY" by Dr. Isadore Rosenfeld. In the article Dr. Rosenfeld points out that the *best way to teach kids healthy habits is to set an example.*

The doctor goes on to say that even though explanations are important, *setting the example for your children is more important than all of the advice, pontifications and admonitions you give them.*

"Do as I say, not as I do" does not impress kids! Warning a child about the hazards of smoking, obesity or lack of exercise will fall on deaf ears if they see you lighting up or stuffing yourself with junk foods as your waistline continues to expand, or living your life as a couch potato.

"The best way to teach kids healthy habits: SET AN EXAMPLE."

THE BODY WAS MADE TO MOVE

"Don't move…you're covered!
First by shadows,
Second by sheets,
And finally and too soon, by shovels."
~ D.M.

"Science may have found a cure for most evils;
but it has no remedy for the worst of them all —
the apathy of human beings."

~ HELEN KELLER

11

"Under the Constitution, a person has the right to die at an early age, but government has the right to urge him not to."

~ PRESIDENT FRANKLIN D. ROOSEVELT

"I want people to realize that you're never counted out. Keep on trying! Live as long as you can; die when you can't help it."

~ JAMES BROWN
Unrelenting Survivor – The King of Soul

"No psychological message is so open to question as that which tells us that we have nothing left to do or to give."

~ ROBERT GRUDIN
"TIME AND THE ART OF LIVING"

The United States ranks 17th in average life expectancy on a list of thirty-three developed nations, according to the U.S. Center for Disease Control.

12 KNOW SWEAT!

SURGEON GENERAL'S
WARNING

INGESTION OF THIS INFORMATION
MAY
BE HARMFUL TO YOUR DOGMA.

14 | KNOW SWEAT!

THE YEAR FOR THE HEALTHY CHILD

The Surgeon General, Vice Admiral Richard H. Carmona, M.D., M.P.H., F.A.C.S., designated 2005 as the *"Year for the Healthy Child."*

The Office of the Surgeon General addressed immunizations, *childhood obesity*, healthy environment, illness and injury prevention, and safe teen driving in an effort to "ensure a healthier population for the next generation."

For more information see:
http://www.surgeongeneral.gov/healthychild/

ON PHYSICAL ACTIVITY AND HEALTH
THE U.S. SURGEON GENERAL'S REPORT

REGULAR PHYSICAL ACTIVITY . . .

- Improves your chances of living longer and living healthier.

- Helps protect you from developing heart disease, high blood pressure and high cholesterol.

- Helps protect you from developing certain cancers, including colon and breast cancer.

- Helps prevent adult-onset diabetes.

- Helps prevent arthritis and may help relieve pain and stiffness in people with this condition.

- Helps prevent the insidious bone loss known as osteoporosis.

- Reduces the risk of falling among older adults.

- Relieves symptoms of depression and anxiety and improves mood.

- CONTROLS WEIGHT.

66% of American adults are seriously overweight or obese *!*

73% of people over sixty-five are overweight *!*

Some doctors require their patients to use railroad freight scales because their office scales only go up to three hundred pounds.

"Over and over, studies show: the fatter you are, the more likely you are to be sick, feel sick and die young."

~ GREG CRITSER
"FAT LAND"

18 KNOW SWEAT!

Some obesity specialists say that if the U.S. obesity epidemic continues to spread at the current rate it won't be long till everyone in the United States is *FAT.* 73% of those over sixty-five years of age are *already FAT.* That's seven hundred-thirty *FAT people* out of every one thousand in one of the fastest growing age groups in our country.

"Be willing to apply your rational thinking to a critical analysis of your own society."

~ *STEVE ALLEN*
"DUMBTH The Lost Art of Thinking"

"If you do not change you can become extinct."

~ *DR. SPENCER JOHNSON*
"WHO MOVED MY CHEESE"

Fat people are jolly folk . . . Santa Claus types. Wrong !

Many Health Care Professionals think the worst result of being overweight is the psychological stress that comes with the package. Being fat is a big disadvantage in our society because it does not fit with our standards of beauty and sexuality. That's why billions of dollars each year are spent on diets, weight loss pills, liposuction, tummy tucks, fat farms and low cal stuff. *There are seventy-one million people in the United States on diets.*

Fat people are paid less than healthy people because they miss more work days and their health insurance is more expensive.

Over one hundred billion dollars a year is spent by the U.S. Health Care System to deal with obesity and related medical conditions.

When scientists looked at the underlying causes of *ALL PREVENTABLE DEATHS* that took place in the U.S. in 2001, they found that together, *lack of exercise and poor dietary habits were the largest underlying cause of death.* Smoking was the second largest. These conditions are worse today.

To paraphrase Garrison Keillor in his "THE BOOK OF GUYS," "*There are millions of folks out there that when they stand up you can hear the tinkle of broken dreams.*"

THE EVOLUTION OF MAN ?...

A FIFTH FREEDOM*!*

Freedom for health and freedom from disease are as vital to America's development as the Four Freedoms enunciated by President Roosevelt. Improved health and physical fitness, in fact, are the basis on which freedom of speech, worship, freedom from want and freedom from fear can achieve full fruition.

"Civilization is a race between education and catastrophe."

~ H. G. WELLS

"The civilized man has built a coach, but has lost the use of his feet. He is supported on crutches, but lacks much support of muscle."

~ RALPH WALDO EMERSON
"SELF RELIANCE"

22 KNOW SWEAT!

OBESITY EPIDEMIC COSTS

The American taxpayers are facing the problem of paying the huge Medicare costs – *one out of every four Medicare dollars* – caused by the obesity epidemic. Their problem is similar to the "Dung Beetle" pushing a ball of elephant dung the size of Marlon Brando up a mountain. Both problems stink and both will break YOU*!*

"More Americans are obese than smoke, use illegal drugs or suffer from ailments unrelated to obesity."

~ Recent RAND UNIVERSITY OF CHICAGO REPORT

"Our *growing softness, our increasing lack of physical fitness, is a menace to our security.*"
~ PRESIDENT JOHN F. KENNEDY

THE ADMINISTRATION ON AGING has materials available to support the Medicare Modernization Act (http://www.cms.hhs.gov/medicarereform/) and help consumers make use of the benefits available under Medicare as of January, 2005. Included in the list of material the AOA has available a brochure from the American Cancer Society (http://www.cancer.org/). American Diabetes Association (http://www.diabetes.org/) and the American Heart Association (http://www.americanheart.org/) entitled Everyday Choices for a Healthier Life (http://www.gov/press/medicare/preventive/EverydayChoicesbro.pdf).

"One way to strengthen Democracy is to strengthen the bodies, minds and characters of those who believe in it."

~ ANONYMOUS

"The problem of conserving national resources is only one part of the larger problem of conserving efficiency. The other part relates to the vitality of our population."
~ PRESIDENT THEODORE ROOSEVELT

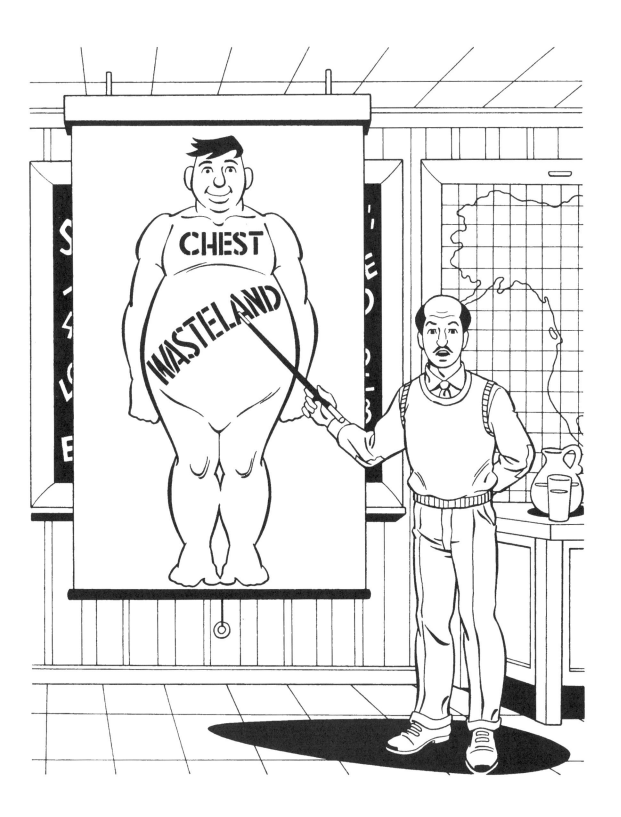

A 2001 study of five thousand students, in thirty-one urban secondary schools, by a group of Epidemiologists from the University of Minnesota researched the daily use of fast food restaurants by students.

THE RESULTS:

A boy that never ate at a fast food restaurant during the school week averaged a daily calorie count of 1,952 calories.

One that ate fast food one to two times a week (more than half of all the children in the study did) during the school week averaged a daily calorie count of 2,192 calories.

Those that ate fast food three or more times a week (20% of those studied) ate a daily average of 2,782 calories.

Researchers pointed out the study showed, "Fast food restaurant use was positively associated with intake of total energy, percent of energy from fat and daily servings of soft drinks."

The kids that ate three or more times a week at a fast food restaurant ate eight hundred more calories a day than the kids that never did eat fast food. That comes to four thousand more calories a week and a disturbing one hundred fifty-four thousand more calories a year.

It takes one hour of vigorous pedaling on a stationery bike or swimming, or jogging, or working out with weights to burn off three hundred–fifty calories.

Considering very few public school systems require Physical Education five days a week, grades 1-12, and that kids sit and watch television or play video games four to seven hours a day, is it any wonder our children are fatter than ever. Diabetes Type II, up to recently a disease afflicting adults, is showing up in our kids.

WELCOME TO THE GOLDEN AGE OF GREASE!

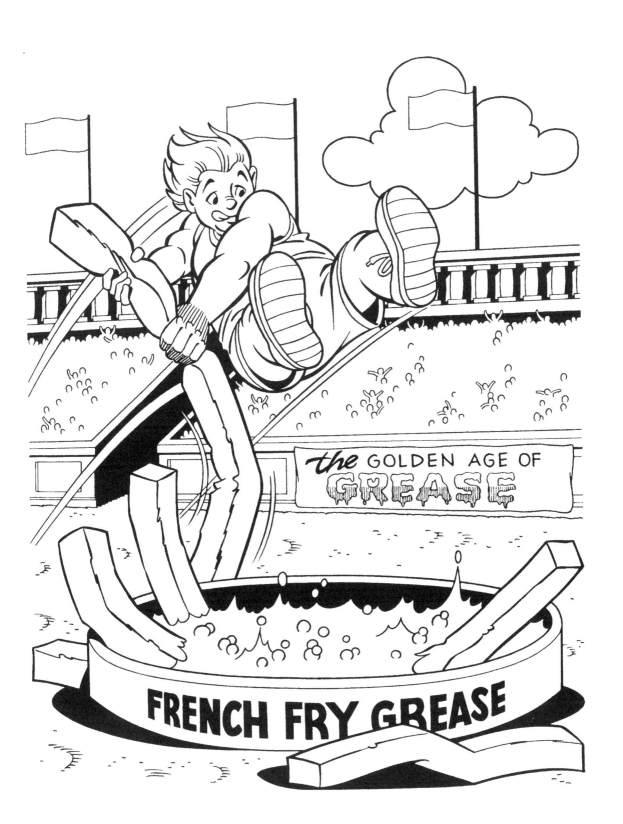

A team of Columbia University researchers reported, by 1993 *forty-one per cent* of all Saturday morning kid show ads were for high-fat foods. The children were being indoctrinated on the benefits of grease, salt and increasing amounts of sugar.

FATAL *is* 60% FAT*!*

What's worse? David Letterman, on Hawaii's victory at the Little League World Series said, "Here's what the kids get. They get free McDonald's and Kentucky Fried Chicken for a year, and fifty-two six-packs of Pepsi. And I'm thinking, well, actually, it might be healthier if they were taking steroids."

PRE-VENTING TOXIC WAIST AT HOME
PREVENT, v. – decisive counteraction to stop something from happening !

- Read a copy of "STRETCHING" by Bob Anderson
- Stretch the muscle groups you are going to exercise before you exercise
- Put exercise equipment in front of the TV
- Exercycle and small dumbells
  KNOW SWEAT*!*
- Use stairs as exercise equipment
  KNOW SWEAT*!*
- Use furniture to help you exercise
- Sit and stand in and out of chair
  KNOW SWEAT*!*
- Use back of chair for support
- Do front, back and side Leg lifts
  KNOW SWEAT*!*

- Use shower and bathtub for therapy
- Stretch in bed or on a soft matt on floor
- Relax
- Don't eat while watching TV except fresh fruit and vegetables
- Drink lots of water

  PARENTS TAKE CONTROL OF
  THE DINNER TABLE

- Don't snack after dinner –
  DON'T EAT AFTER DINNER
- Don't ever skip breakfast –
  BREAK-FAST

  This is a GOOD START*!*

28 KNOW SWEAT!

APPLES AND PEARS

People put on excess body weight in two locations. Apple shaped people add excess FAT around their waist. They are at a greater risk for certain cancers – colon, breast and prostate, diabetes and heart disease.
TOXIC WAIST!

Pear shaped people add their excess FAT around their hips making them at risk for joint problems and other ailments.

You don't see many very old Apple and Pear Shaped people. *They die younger than Fit People.*

KNOW SWEAT! 31

"Bring me a FAT man," a physician told the New York Times, "and I'll show you a diabetic, or someone who will become one."

Diabetes is an insidious disease. Other than a large waistline, symptoms are rare. Diabetics run the risk of cardiovascular disease, kidney failure, high blood pressure, stroke, blindness and amputation. "Our 'Diabetes Epidemic' reflects an aging population that is increasingly overweight and sedentary."

~ TREVOR ORCHARD, M.D.
Lead Investigator Diabetes Prevention Program

Thomas Herrion was a National Football League Player. He was a happy, friendly guy. He was well liked by his teammates and coaches. He was a playful player who would sometimes break into song. He was enjoying the game of football and feeling good about himself for the contributions he'd made in the pre-season football game he had just been in. He had played lineman in the last series of downs in the game.

After the game he was making the rounds of players and coaches from both teams, shaking hands, slapping backs, hugging teammates – always smiling. Thomas Herrion smiled a lot.

In the locker room the team and coaches knelt in prayer as they did after every game. There was a commotion at the back of the room. Thomas had collapsed. In spite of all the professional emergency aid right there in the locker room, Thomas died. He was twenty-three years old. He did not drink or smoke.

Thomas Herrion will be missed by his family, friends, teammates and his coaches.

Thomas was six feet-three inches tall and weighed three hundred-fifteen pounds. On the day of his death there were *three hundred-seventy five NFL players that weighed over three hundred pounds.* Only one of the thirty-two NFL teams does not have a line that averages three hundred pounds or more. *These days it is not unusual for college teams to "boast" of lines averaging three hundred pounds.* We are seeing more three hundred pounders in high school. Because of Thomas' death and the death of other athletes, the NFL has committed to looking into the early deaths and their possible causes.

BEING FAT IS
MORE DEADLY
THAN SMOKING

KNOW SWEAT! 33

"I've stopped eating chili dogs.
I gave up coffee and cigars.
I'm all right.
I miss the taste of a good short triple latte.
But you do these things if it makes the difference between living and dying."
~ BILL COSBY
"I AM WHAT I ATE...and I
AM FRIGHTENED!!!"

"Illness is telling us what we need to stop doing. If we look at illness that way, then it has great value."

~ O. CARL SIMONTON
"FOCUS"

Exercise prolongs life...THE GOOD LIFE

According to an ongoing study into sexual activity among the aging, researchers at Duke University found that 80% of men in their late sixties continue to be interested in sex...*exercise.*

Exercise in general can improve your sense of well-being. People that exercise sleep better. In a study at Fred Hutchinson Cancer Research Center, women ages 50 to 75, who started taking brisk, half-hour morning walks improved their ability to fall asleep at night by 70%.

Weight training increases lean muscle mass which helps control a person's weight, increases strength and helps with balance.

So, *exercise to live.* Besides, it has been illegal to die of old age since 1951, when the federal government deleted *"old age"* from the official list of *"causes of death."* Does that mean if we live a healthy, active life, to a ripe old age and die of natural causes...we can bow out, breaking the law, and never get caught?

37

Ten years ago the average dress size in America was a size 12. *Today it is a size 14.*

There are several women's specialty stores across the nation (one chain has more than two hundred stores) that sell sizes 14 to 28 only – *no dresses sold under a size 14.*

Let's not leave the men out. A very successful higher-end department store, known for customer service, confronted an unexpected problem with a manufacturer's tag describing the cut of their trousers. The manufacturer offers *"Traditional,"* *"Classic," "Full," and "Easy" fit.* Enough customers complained that *"they were being singled out by the FULL tag"* and to accommodate the customers, the men's departments removed the *"FULL"* tags from the slacks.

All this time we have been wrong. The map *IS the territory.*

> **It is quality rather than quantity that matters.**
> **~ SENECA**

Don't judge a *look* by it's cover *!*

X-HER-SIZE

EXERCISE

"Human kind cannot bear very much reality."

~ T. S. ELIOT

The more serious the illness, the more important it is for you to fight back, mobilizing all your resources — spiritual, emotional, intellectual, *PHYSICAL.*

~ NORMAN COUSINS
"ANATOMY OF AN ILLNESS"

HOW MANY
HANDLES
ON YOUR
GIRTHDAY
CAKE?

KNOW SWEAT! 41

"Prioritize exercise in your life. It is just too powerful a fat fighter for you not to have on your side. There is absolutely no way you can control your weight for a lifetime without it."

~ DR. PHIL McGRAW

Studies show that people keep at an exercise program longer if they exercise with a friend.

"LET'S GET PHYSICAL"

~ Sung by *OLIVIA NEWTON-JOHN*

The main reason people give up exercising is that they want to be fit, in shape *TOMORROW.* They have been lied to for many years . . . *"exercise on this piece of equipment for just six minutes a day, three times a week and you will be in shape."* WRONG*!*

Most exercise contraptions concentrate on a specific muscle group – inner thigh, buttocks or abs. But what about the rest of your muscles? Some exercise machines are versatile and you can get a work out using one of them. These machines use large rubber bands or stacks of metal weights or your own body weight (hydraulics or cords and bows) for resistance. These machines are excellent for home or gym work outs.

The individual that wants to be fit by tomorrow, in six minutes, will exercise too hard, too fast, too much, too soon, no matter what they use for resistance. They abuse their muscles and the result is pain, stiffness, swelling and sadly, injury. All are used for an excuse to give up exercising.

"ME TARZAN, YOU TRAIN*!*"That was the name on my brochure when I was a personal trainer. I told my clients "Slow down*!* Don't lift that much weight*!* Rest between exercises*!* Be kind to yourself*!* Don't hold your breath when you lift*!*"

To quote Dr Phil McGraw in his book, "THE ULTIMATE WEIGHT SOLUTION — THE SEVEN KEYS TO WEIGHT LOSS FREEDOM," "One of the worst things I see with people trying to lose weight (get Fit, get IN SHAPE) is that they become obsessive about exercise. So much so, that this concern rises to a level of disorder, self-defeating behavior that can inflict a great deal of physical and psychological damage."

Other books to read for effective, safe successful workouts are:

"STRONG WOMEN AND MEN BEAT ARTHRITIS"
    by Miriam Nelson, PhD, Associate Professor of Nutrition at Tufts University
"GET WITH THE PROGRAM"
    by Bob Greene.
"BODY BUILDING FOR MEN" and "BODY BUILDING FOR WOMEN"
    by Governor Arnold Schwarzenegger

"Where are all the Atkins Alumni?" *asks Jules Hirsch, M.D., Obesity Researcher and Professor Emeritus at Rockefeller University, New York.* He reports that most of the ex-Atkins dieters he has seen, have gained all their lost weight back. August, 2005, Atkins declared bankruptcy*!*

"When the will to act is thwarted it atrophies into a wish to be taken care of."
~ ELLEN J. LANGER
"MIND-FULNESS"

"*Repeated failure (to lose weight) brings about* 'learned helplessness.' *Even when solutions are available*, a mindless sense of futility prevents a person from reconsidering the situation. *If we look for new ways to deal with our over-weight problem we probably could prevent learned helplessness.*"
~ DR. MARTIN SELIGMAN
Research Psychologist

"A major source of my own mental turmoil is the news. *The percentage of stories that make me feel anxious or outraged is very large and increasing, as news media focus more and more on murder, mayhem, and misery. It is easy to forget that we have a choice as to whether we let this information into our minds and thoughts. I recommend* 'NEWS FASTS.' *I think these fasts will allow you to get better rest and sleep.*"

~ ANDREW WEIL, M.D.
"SPONTANEOUS HEALING"

The doctor suggests we try an Eight Week News Diet. *The first week do not watch, read or listen to any news for a day and see how you feel. The next week try two newsless days. The next five weeks add another day of no news until you go a whole week on the news diet. At the end of the week think about how much news you want to let back into your life.*

What a great opportunity to spend those "bad news minutes" enjoying a fun, physical activity of your choice. THAT'S A DIET THAT WORKS*!*

Walter Willett, M.D., and his fellow researchers at Harvard Medical School and The Harvard School of Public Health have built a "Healthy Eating Pyramid." It is the basis of their book "EAT, DRINK AND BE HEALTHY." The "Healthy Eating Pyramid" is built on a foundation of *daily exercise and weight control.*

Their "Healthy Eating Pyramid" is solid. It is held together by evidence from many kinds of research. It is an eating lifestyle based on science. It is a multicultural approach to healthy eating backed by the latest food studies from around the world.

If used, the information in Dr. Willett's book has two payoffs. One: its section on recipes will make eating a healthy pleasure. Two: if the eating plan is followed, it will be an important part of protecting yourself against a long list of common diseases – including heart disease, stroke, several common cancers, and diabetes.

When you make this healthy diet a habit and combine it with regular exercise (and eliminate smoking) you can reduce heart disease 81%; stroke and some cancers by 70%. *This could be the single most important tool for improving your health and the health of our nation.*

Dr. Willett and his colleagues at Harvard University are immune to influences by special interest groups. They have gathered all their research with a goal of improving our health. They are not in the business of selling any foods or food groups. The USDA's old Food Pyramid compiled by the U.S. Department of Agriculture and the U.S. Department of Health and Human Services has been much improved, since April, 2005, when the U.S. Department of Agriculture Center for Nutrition Policy and Promotion came out with their new Pyramid version – www.mypyramid.gov.

Read Dr. Willett's book.

"Gluttony is not a secret vice."
~ ORSON WELLES

Here are the latest additions to the menus out there in Fast Food Land...

- The Monster Thick Burger = 1,420 calories; 540,000 sold the first week.

- The Belly Buster – start with a large portion of hash brown potatoes, topped with eggs, meats, veggies, and cheese, a side of pancakes, plus 2 strips of bacon and 2 sausages.

- A stack of three meat patties (three hamburgers in one sandwich) with all the goodies added.

- You can get the Westward Ho Mega Dog on the Las Vegas Strip. It is fourteen inches long and weighs three-quarters of a pound.

- Throughout the state of Texas there are billboards advertising a FREE steak if you eat *ALL 52 ounces.*

The above "epicurean delights" were advertised in local newspapers and on television.

You can now buy MEGA M & Ms *!* They are 55% bigger than the standard M & Ms *!*

Bill Cosby, in his wonderful book "I AM WHAT I ATE...AND I'M FRIGHTENED *!!!*," tells of recently watching a teenager eating a chili dog with sauerkraut and gulping down a soft drink while, at the same time, wolfing down handfuls of French fries covered with cheese and ketchup. Get his book and read his comical observations of this scene.

He closes by writing, *"I think to myself: he is what he's eating and about forty years from now he's going to be frightened."* Mr. Cosby has been teaching us with his comedy for a long time.

ORSON WELLES' DOCTORS WARNED HIM TO GIVE UP EATING DINNERS FOR FOUR UNLESS THREE OTHER PEOPLE WERE EATING WITH HIM ...

HE DIDN'T LISTEN!

GLUTTON, n.

"A person who escapes the evils of moderation by committing dyspepsia."

~ AMBROSE BIERCE
"THE DEVIL'S DICTIONARY"

EATING RECORDS . . . *Names withheld to protect the guilty!*

Over 8 pounds of baked beans in under 3 minutes...*Ain't that a gas!*
65 hard boiled eggs in less than 7 minutes...*Egg-straordinary!*
Over 1$^{1}/_{2}$ pounds of salted butter in 5 minutes...*Buttered butter!*
Over 3 pounds of beef tongue in 12 minutes...*Hold your tongue!*
20 pounds of rice balls in 30 minutes...*Have a ball!*
5 pounds of birthday cake in less than 11$^{1}/_{2}$ minutes...*Happy Girthday!*

There are more than seventy-five competitive eating records *sanctioned* by the International Federation of Competitive Eating (I.F.O.C.E.).

SANCTIONED, n.

"Support or encouragement, as from public opinion or established custom."

~ *THE AMERICAN HERITAGE DICTIONARY*

"A CULTURE THAT CONDONES OBESITY, WHETHER CONSCIOUSLY OR UNCONSCIOUSLY UNDERMINES ANY ATTEMPT TO CONVINCE PEOPLE TO LOSE WEIGHT."

~ POLICY ANALYSIS, Inc. (PAI) STUDY

53

TO WELL WITH YOU

Bob Greene authored the #1 New York Times Best Seller, "GET WITH THE PROGRAM—Getting Real About Your Weight, Health and Emotional Well-being." This book is a great resource on *"how to"* start and continue a progressive resistance exercise routine. Greene lays out individualized exercise programs that will help make a person stronger and leaner.

I've done an exercise routine similar to the ones listed in this book, most of my life. And, at age seventy-one, I can say that *exercising has given me a better quality of life.* I have more confidence because I have more strength. *It has helped control my weight, reduced anxiety and helped me win the battle against Clinical Depression.*

I have a degree in Physical Education, from University of California, Los Angeles, and agree with Bob Greene — *regular exercise is one of the most important things you can do for yourself throughout your life.*

55

TOXIC WAIST?... Get To KNOW SWEAT*!*

"We're just too darn fat, ladies and gentlemen, and we are going to do something about it !" said Tommy Thompson, Secretary, U.S. Health and Human Services. Thompson recently introduced a new, multimillion dollar ad campaign that encourages people to take *"small steps"* toward controlling their children's diets and encouraging more physical activity, such as climbing stairs instead of taking the elevator.

~ LAS VEGAS REVIEW-JOURNAL

As we age, excess weight causes and complicates bone disease. One pound of extra body weight puts from two to four pounds of additional stress on the knees and hips. So, if you are ten pounds overweight your knees and hips are carrying an extra forty pounds of extra pressure, even during routine movement…fifty pounds overweight and those joints grind under one hundred to two hundred pounds of unnecessary pressure. If we dance, jog, or go up and down stairs the pounds of pressure increase dramatically.

Here comes Osteoarthritis. Cartilage wears away cartilage and exposed bone surfaces grind against each other, thus causing swelling, pain, and big problems getting your body from Point A to Point B. Studies show that if the average person loses ten pounds he will cut Osteoarthritis pain in half. To dance or walk or just move your body away from the refrigerator with half the pain seems like a pretty good idea.

~ GREG CRITSER
"FAT LAND"

Want to get stronger, have less pain from arthritis? Read "STRONG WOMEN AND MEN BEAT ARTHRITIS" by Miriam Nelson, Ph.D., Assistant Director of Nutrition, Tufts University. Dr. Nelson and her colleagues have developed a strengthening routine to help you beat the aches and pains of Arthritis and Osteoporosis. The exercises will help control weight gain. *Muscle is the only calorie burning engine your body has. Lose muscle – gain fat!* Her book points out that you don't have to join a gym. Nelson and her team conducted an at-home strength-training program and reported a 43% decrease in pain and a 71% increase in muscle strength in just sixteen weeks. Get her book and follow her routine with your doctor's okay.

THAT'S A KNOW-KNOW *!*

One of the best definitions of *"learning"* is: *"that process of mind that allows us to take in new information, retain it, and make a permanent change in out potential behavior."*

That means we can learn how to do something, learn the answer to a problem, and still not demonstrate it *(solve it)*.

Almost everyone knows that exercise is very good for us. Why don't **66%** of Americans do it? Exercise *!*

If you are one of the seven out of ten citizens that do not exercise, let me ask you "Is your give a damn broken?"

If you don't care about living a joyful, productive life, and you don't care that you are a lousy role model for your kids...
HOW SAD IT THAT !

"Perhaps the most valuable result of *all* education is *the ability to make yourself do the thing you have to do*, when it ought to be done, whether you like it or not."

~ THOMAS HENRY HUXLEY

KNOW SWEAT!

Do you know anyone that has been on the *FAT/THIN Roller Coaster?* Some people have been riding it for years. They go on a strict diet for thirty days – the latest *FAD DIET* – *say they just eat dandelion greens and drink water for thirty days, 750 calories a day.* What happens? Of course their weight goes down. *They lose fat but they also lose muscle. Their metabolism also goes down. Their body is no dummy... "Hey, the boss is going on another diet, starving us for a month. So, shut down the metabolism; hold on to our fat reserves."*

The diet ends – they all do. Here come the calories and UP goes the weight. *The* "weight roller coaster" *goes up faster and higher now that you are back on a pre-diet eating plan. Why?...more calories in, less muscle to burn them.*

And so it goes. Then next time we go on a diet, and we always do, it is harder to go down in weight and easier to go up.

The result...*each diet we end up with less muscle – the body's biggest burner of calories – and more fat. NOT GOOD.* This *"weight roller coaster"* is not fun. You hear dieters say, "I've lost nine hundred pounds, the same thirty pounds thirty times*!*"

This is very discouraging and also their body fat ratio to lean muscle mass tilts toward FAT. Bad for the heart. Very bad for the self-confidence, self image, and opens the door for all kinds of serious medical problems.

"Probably nothing in this world arouses more false hopes than the first day of a diet." says Dan Bennett.

DIETS DON'T WORK*!*

CALM UP

Want to get rid of a bad habit? Want to overcome a fear? Want to learn something new? You're not alone.

Here is a tool you can use to help you do all of the above. It's called *"The Relaxation Response."* Dr. Herbert Benson, Harvard Medical School wrote the #1 Bestseller *"THE RELAXATION RESPONSE."* He also wrote *"BEYOND THE RELAXATION RESPONSE"* and *"YOUR MAXIMUM MIND."* He's been doing research on the mind for almost half a century.

It's a simple relaxing routine and it doesn't cost a cent. You can do it anywhere but it works best in a quiet place. I've taught this very easy way to do it at the Center For Healthy Living, at Eisenhower Medical Center, Rancho Mirage , California. *IT WORKS!*

It goes like this...
1. Find a quiet place. It can be your favorite chair (inside or outside). If a friend or relative joins you, he/she has to be quiet also. If there is conversation it can't be about money, politics, or the state of your teeth.
2. Pick a focus word or a short phrase that's firmly rooted in your belief system. A religious person might choose a line from a favorite prayer or hymn. The words *PEACE or CALM* work for some people.
3. Sit quietly in a comfortable position.
4. Close your eyes.
5. Relax your muscles.
6. Breathe slowly and naturally, and as you do, silently repeat your focus word or phrase while you exhale.
7. Assume a passive attitude. Don't worry about how you're doing. Thoughts will stray into your mind. Just ignore them and gently go back to your repetition. It is *NOT a concentrated effort. It IS an effortless concentration.*
8. Continue for ten to twenty minutes, once or twice a day.

This will slow the chaos in your mind, *Brain calming* instead of *Brain storming. It opens your mind.* That's Phase A.

Phase B – Now, with your mind open, expose your mind for at least fifteen to twenty minutes to important influences that will help renew your mind with thoughts of reaching your goal.

If you are aiming at a healthier lifestyle, plan an exercise activity and then do it. If your problem is overeating, make a vow NOT to do it. Read a book on the subject to find new ideas. Discuss it with your doctor or friends or relatives.

Sound simple? It is and it works. *THINK SMALL!*

Weight training strengthens muscles, making movement easier and less painful. In studies, seventy year old subjects that lifted weights for several years had the same amount of muscle as twenty-eight year olds.

"There are studies on people close to one hundred years old who have gotten stronger through resistance training." says Kent Adams, Ph.D., Associate Professor of Exercise Physiology at the University of Louisville.

In an eight week study, nursing home residents ages eighty-six to ninety-six increased their muscle strength by 174% and increased their walking speed by 48%.

A University of Southern California study of seventy year old men showed increased muscle strength after just eight weeks of strength training.

"SUCCESS COMES IN CANS
NOT IN CAN'TS."

~ DR. MARILYN GELLIS
"MENTAL FLOSS"

"Headed, I fear, toward a most useless place…
THE WAITING PLACE !"

~ DR. SEUSS
"OH, THE PLACES YOU'LL GO !"
(one of forty-four books by Theodor Geisel)

"My life is one long obstacle course with me as the chief obstacle."
~ JACK PAAR

"Fear is that little dark room where negatives are developed."
~ DR. MARILYN GELLIS
"MENTAL FLOSS"

ATTITUDE

OUR ATTITUDE IS UP TO US

"The capacity for happiness may be as important as the opposable thumb in explaining the success of the human species. It is important for motivational reasons that people not be in a negative mood most of the time. An optimistic state of mind is a prerequisite to obtaining such goodies as food, shelter and social support. Positive moods may motivate human sociability, creativity, and produce a strong immune response to infections."

~ NEWSWEEK, 1996

GET OFF THE COUCH.

You don't get to chose how you're going to die or where. You can only decide how you're going to live – NOW *!*

~ DR. MARILYN GELLIS
"MENTAL FLOSS"

IT'S NEVER TOO LATE OR TOO SOON
TO START A HEALTHY LIFE STYLE !

MOTIVATION to exercise, to choose a healthy life-style?
Use whatever works for you.

HATE . . . You hate the way you look and feel.

FEAR . . . The fear of not being able to move, to get upstairs, to carry a child, to get around, to dance, to skip, to run, to walk comfortably! The fear of living your life in a "phone booth."

GUILT . . . The guilt of not being able to help a loved one in an emergency. The guilt of being a "lousy" role model for your children and grandchildren. The guilt of dying without having lived.

GREED . . . Being able to hoard all the rewards that come with being fit. Delighting in the joy of movement. Holding on to the happiness that comes with the freedom from the chains of illness.

PRIDE . . . The child-like feeling...I can do it myself. Look at me...I am strong, agile, limber, enduring enough to do it myself.

SEE . . . I am showing you my Personal Declaration of Independence!

Whatever moves you, use it *!!!*

ABOUT EXERCISING

No Pain…No Gain… WRONG*!*
That is for athletes – professional, amateur,
Olympic and weekend athletes.

NO PAIN…ALL KINDS OF GAIN
THIS IS FOR 99% OF US.

*Better body weight control, better balance, more strength,
more range of motion, more agility, more endurance, more
confidence, more relaxed, more fun. These gains are ALL
possible WITHOUT PAIN. Pain is a warning. If the exercise
hurts, you have gone too far too soon. If the exercise hurts,
back off, slow down, do less.*

If your goal is to jump higher, run faster, or lift more weight than everyone in the world, then you will have to deal with the "pain barrier." Pain will be your buddy every work out.

But for the 99% of us that just want to be fit and healthy, going for the burn, exercising a muscle group to failure, dealing with pain is the surest way, the most powerful excuse people use to stop exercising.

Why do you think that most athletes are walking ads for Johnson and Johnson? *When you ignore the body's pain warning system, injuries will happen.*

"Let's see…it's time for my daily exercise. Do I want to torture myself for an hour? *DUH!*"

Make exercise fun or you won't exercise. It is that simple.

You said, *"but."*
I've put my finger on the whole trouble.
You are a *"but"* man.
Don't say, *"but."*
That little word *"but"* is the difference between success and failure.
Henry Ford said, "I'm going to invent the automobile," and Arthur Flanken said, *"But…"*

~ SGT. ERNIE BILKO
THE PHIL SILVERS SHOW

GET OFF YOUR BUTT*!*

"Some people regard discipline as a chore. For me, it is a kind of order that sets me free to fly."
~ JULIE ANDREWS

"Exercise and temperance are the secrets by which our youthful *vigor is carried over into old age.*"

~ *CICERO*

The value of exercise is not really a new concept. Cicero lived 2,100 years ago!

Don't have time to exercise? Oh…then make time for heart disease, high blood pressure and high cholesterol (the bad kind).

75

Seventy-eight million Baby Boomers, men and women, are now between the ages of forty-five and fifty-nine.

In the July/August, 2005 issue of "AARP The Magazine" there is a list of celebrities over the age of fifty that have become, to quote The Magazine, "Sexier with seasoning." *Their sex appeal is getting stronger with age.* Take a look at the list of sixty-seven men and women. You will find it hard to find *just one* of them that is over weight.

"Self respect is impossible without respect for one's own past, and intelligent action in the present is impossible unless it is based on a plan that extends into the future."

~ *ROBERT GRUDIN*
"TIME AND THE ART OF LIVING"

Cherish your health. If it is good, maintain it. If it is unstable, improve it. If it is beyond what you can improve, *get help!*

ENROLE IN THE SCHOOL
OF APPLIED PARTS

The Journal of Physical Education, Recreation and Dance announced that Thomas L. McKenzie is the recipient of the R. Tait McKenzie Award for 2005.

Professor Emeritus of Exercise and Nutrition, San Diego State University, Thomas McKenzie is a leader for research and program development in Physical Education. His landmark research and curricular projects – CATCH (Child and Adolescent Trial of Cardiovascular Health), SPARK (Sport, Play and Active Recreation for Kids) and M-SPAN (Middle School Physical Activity and Nutrition) are known world-wide. Over seven hundred elementary and middle schools in the United States now use his evidence-based physical education programs to guide children toward a lifetime of physical activity.

Many dedicated Physical Education Teachers, like Professor McKenzie, are out there in the trenches fighting a continual battle of school budget cuts, parents disconnect between schools and their child's physical fitness level, bad diets including too much sugar and other "junk foods" and an army of couch potatoes spending four to seven hours a day watching television and video games instead of playing "Fun" games.

These Physical Education, Recreation, Dance, and Health Teachers should be congratulated for their efforts in a very steep, up-hill battle.
THEY KNOW SWEAT!

"Life is a beautiful thing as long as I hold the string. I'd be a silly so-and-so if I ever let go." says Dr. Wayne Dyer in "PULLING YOUR OWN STRING."

"Add variety – choose three or four physical activities you enjoy and you'll enjoy learning to exercise even more. NEVER BORED!" says Dr. Phil .

"Book reading is a solitary and sedentary pursuit, and those who do are cautioned that a book should be used as an integral part of a well-rounded life, including a daily regimen of rigorous physical activity, rewarding personal relationships, and a sensible low-fat diet. A book should not be used as a substitute or an excuse."

~ GARRISON KEILLOR
"THE BOOK OF GUYS"

PRE-VENTING TOXIC WAIST IN SCHOOL
PREVENT, v. – decisive counteraction to stop something from happening!

- Parents demand daily Physical Education classes taught by Physical Education Teachers
- Parents demand healthy menus in the school cafeteria
- Parents ask for bottled water machines to replace soft drink machines on campus
- Parents ask for healthy snacks to replace junk-food in campus vending machines

- Walk to school – walk home   when safe and weather permits
- Participate in school sports – Intramural activities available for ALL students. Support school athletic teams by participating in cheerleading, marching band, flag and banner waving, baton twirling and attending games
This is a GOOD START!

LOOK YOUR AGE

"You don't look sixty or forty or seventy-five*!*" Translation…*You look much younger than the age you say you are.*

But you really do look sixty or whatever age you are. You have just taken care of yourself. You have exercised all your life, watched what you ate and didn't eat, got sufficient rest and enjoyed your work so you look how a fifty or sixty or seventy year old should look.

Most people in our society haven't lived a healthy lifestyle and so they look ten to fifteen years older than they really are.

*   *   *   *

"Albert Schweitzer once said disease tended to leave him rather rapidly because it found so little hospitality inside his body."

~ NORMAN COUSINS
"ANATOMY OF AN ILLNESS"

THE GOAL

"You lose weight for yourself, not others – *you choose a realistic weight as a goal and think about how you will feel when you reach it.*"

~ DR. PHIL McGRAW

"Admit that there is *no PERFECTION, but in each category of achievement something is Tops; be satisfied to strive for that.*"

~ DR. HANS SELYE

"*Success is peace of mind which is a direct result of self-satisfaction in knowing you have done your best to become the best you are capable of becoming.*"

~ COACH JOHN WOODEN

"*We do not want our children to become a Nation of Spectators. Rather, we want each of them to be a participant in the Vigorous Life.*"

~ PRESIDENT JOHN KENNEDY

83

IT'S COOL TO SWEAT

"When our body is exposed to heat, we sweat and the evaporation of perspiration from the surface of our skin has a cooling effect."

~ DR. HANS SELYE
"STRESS WITHOUT DISTRESS"

YOU CAN BANK ON IT

"Imagine you had a bank that each morning credited your account with $1,440 – with ONE condition: whatever part of the $1,440 you failed to use during the day would be erased from your account, and not carried over. What would you do? You'd draw out EVERY cent EVERY day and use it to your best advantage.

Well, you do have such a bank, and it's name is TIME. Every morning, this bank credits you with 1,440 minutes. And it writes off as forever lost whatever portion you have failed to invest to good purpose."

~ DR. MARILYN GELLIS
"MENTAL FLOSS"

PRE-VENTING TOXIC WAIST AT WORK
PREVENT, v. – decisive counteraction to stop something from happening !

- Take the stairs to and from your office.
- Take a break every hour. Studies show people that take periodic breaks are more productive
- Actually drink water at the water cooler
- If your company has an exercise facility – use it.
- On a business trip, use the hotel's pool and gym
- Brown bag healthy snacks (Fresh fruits and vegetables, nuts or health bars and drinks) or purchase them at work
- At lunch – take a stroll, walk on a beach or in a sand-box (bare foot), spend time in a park, relax, people watch, listen to music, wear earphones to blot out the sounds of the city, day-dream, re-create, plan a vacation, enjoy the company of a friend and tell them you won't talk about politics, money, religion or the state of your teeth

- Enjoy your meal *!*
- If your work is very physical and you still have TOXIC WAIST, take a serious look at your eating habits. Make some healthy choices like:
  - Eat less refined sugars, salt, and saturated fats
  - Eat more fresh vegetables, fruits, nuts and legumes
  - Drink lots of water, not soft drinks
  - Cut the size of your servings down
  - Don't eat after dinner and never miss breakfast, a healthy breakfast
  - Break your over night fast and eat some energy for the start of your day

  This is a GOOD START *!*

Nike has teamed up with *SPARK (Sports, Play Active, Recreation for Kids)* and HEAD START to form *NIKE GO HEAD START.* It is a program for preschool children *and* their parents that has a goal of increasing the physical activity of preschoolers. It will also encourage parents to be active role models in raising healthy, fit children. The program will send information to families about fun activities that can increase movement.

The Pilot Program will train three hundred-twenty teachers and with donated equipment, they will give thirty thousand kids *a head start in making physical activity a fun habit for life.*

It is a very important project to help make the country's obesity epidemic slow down and eventually go away.

Like the title of Anthony Newley's song, *"Educate The Parents Before It's Too Late!"*

Read Greg Critser's "FAT LAND How Americans Became the Fattest People in the World" for more information; or call Nike.

The State of Nevada recently passed Bill 313 called the "Common Sense Consumption Act," sponsored by Senator Dennis Nolan, R-Las Vegas. It bans obesity related lawsuits aimed at fast food restaurants. Seventeen states have already enacted such legislation. Shortly afterword, the United States Congress passed a similar bill making it a Federal Law. Politicians unofficially call it "The Cheeseburger Bill."

"Common Sense Consumption Act" is a good name for this Bill.
What we eat and don't eat is *OUR RESPONSIBILITY.*

Our total health is our responsibility!

RESPONSIBILITY, n.
"A detachable burden easily shifted to the shoulders of God, Fate, Fortune, The Devil, Luck or one's neighbor."

~ AMBROSE BIERCE
"THE DEVIL'S DICTIONARY"

Your Level of Fitness is Your Personal Declaration of INDEPENDENCE!

Two battleships assigned to the training squadron had been on maneuvers in heavy weather for several days. The visibility was poor with patchy fog, so the Captain remained on deck. Shortly after dark, the lookout on the wing of the bridge reported, "Light bearing on starboard bow."

"Is it steady or moving astern?" asked the Captain. And the lookout replied, "Steady, Captain." Which meant they were on a dangerous collision course.

"Signal that ship: We are on a collision course, advise you change course 20 degrees." said the Captain. Back came a signal, "Advisable for you to change course 20 degrees."

The Captain said, "Send, I'm a Captain, change course 20 degrees." And the reply was "I'm a Seaman Second Class. You had better change course 20 degrees."

By that time the Captain was furious. He spat out, "Send, I am a battleship. Change course 20 degrees." Back came the flashing light, "I'm a lighthouse." The battleship changed course.

That little bit of information let the Captain know he and his crew were headed for an untimely death. What he had thought was real but couldn't see through the fog was a lie he'd told himself.

Hopefully, with the information in this book about our country's "*obesity epidemic*" gathered in scientific studies by health professionals, physical education teachers, Exercise Physiologists, Medical Doctors, Psychiatrists, Professors, Nutritionists, Child Care Research Specialists, Gerontologists and Pharmacologists *will motivate us as individuals and as a nation to change our course.*

"The health, safety, and physical education program in the elementary and secondary schools and in higher institutions should be given a high order of priority among competing educational interests."

~ *Educational Policy Commission,*
National Education Association

This was one of the goals of the N.E.A. sixty years ago. What happened?

**CALIFORNIANS HAVE GAINED
360 MILLION POUNDS
IN THE LAST TEN YEARS.**

One in three children in
California are overweight and 40%
of school-aged children are unfit.

The economic burden of physical inactivity, overweight, and obesity in adults was twenty-two billion dollars for California in 2000, and was expected to rise to twenty-eight billion dollars in 2005.

California is not the only State that has an OBESITY EPIDEMIC. Our whole country is under the health threatening and economic pressures of an OBESITY EPIDEMIC.

This is an unacceptable state of affairs to California Governor Arnold Schwarzenegger and First Lady Maria Shriver. So they called for a "GOVERNOR'S SUMMIT on HEALTH, NUTRITION and OBESITY." The Summit was sponsored by The California Endowment, headed by CEO Dr. Robert Ross.

The following message is a press release from The Governor, The Secretary of California Health and Human Services, Kim Belshe and The California Endowment:

If California is to lose the three hundred-sixty million pounds it has gained over the last decade, bold action is necessary – by individuals and families, as well as business, community and government leaders – to promote an environment that encourages healthy eating, regular physical activity and responsible individual choices. To that end, Governor Arnold Schwarzenegger will set forth the following 10-point vision for a healthier California. The Governor recognizes that this comprehensive long-term vision will not be realized overnight. Rather, he has outlined an ambitious, forward-reaching guide to challenge all of us. – government, business, community organizations and individuals – to make California a national model for healthy living.

A VISION FOR CALIFORNIA
10 STEPS TOWARD HEALTHY LIVING

1. Californians will understand the importance of physical activity and healthy eating, and they will make healthier choices based on their understanding.
2. Everyday, every child will participate in physical activity.
3. California's adults will be physically active everyday.
4. Schools will only offer healthy foods and beverages to students.
5. Only healthy food and beverages will be marketed to children ages 12 and under.
6. Produce and other fresh, healthy food items will be affordable and available in all neighborhoods.
7. Neighborhoods, communities and buildings will support physical activity, including safe walking, stair climbing, and bicycling.
8. Healthy foods and beverages will be accessible, affordable, and promoted in grocery stores, restaurants, and entertainment venues.
9. Health insurers and health care providers will promote physical activity and healthy eating.
10. Employees will have access to physical activity & healthy food options.

At The Summit Governor Schwarzenegger not only unveiled his VISION FOR A HEALTHY CALIFORNIA but also took immediate action to ensure California's students have healthy food and beverages in schools. On September 15, 2005, he signed Bills sponsored by Senator Martha Escutia. These Bills will give California's schools the strongest nutrition standards in the nation. He also signed another Bill sponsored by Senator Abel Maldonado, which will provide money in the Governor's budget to include more fresh fruits and vegetable in the school meal program.

By helping the State of California take steps to correct it's *OBESITY EPIDEMIC*, Governor Arnold Schwarzenegger has created a national model for our country to follow.

"Altruistic egotism. It is the only way to preserve teamwork whose value is ever increasing in modern society. It works on the cellular level up through cooperative 'mental insurance' groups such as the family, tribes and nations within which *altruistic egotism* is the key to success."

~ *DR. HANS SELYE*
"STRESS WITHOUT DISTRESS"

"In other words, *life is a team sport.* An obesity epidemic can be beaten by teamwork."

You help me, I'll help you. A weight problem? You help me help myself and vice versa and we'll beat it. That's how it works.

~ *D.M.*

PRE-VENTING TOXIC WAIST IN THE COMMUNITY
PREVENT, v. – decisive counteraction to stop something from happening !

- Ask city officials to provide safe, well lit walking, jogging and bicycling trails
- Join and attend: swimming, dancing, hiking, biking, walking, tennis, exercise and Tai Chi classes & clubs, or form your own group with friends
- Attend local sporting events: team sports, charity walks, jogs and biking events
- Attend Health Fairs
- Volunteer for road cleaning groups
- Get involved with Scouting — Boy — Girl — Sea Scouts
- And the Boys and Girls Clubs of America
- Use available facilities in Recreation Parks and School Playgrounds, soccer fields, basketball courts, tennis courts, swimming pools
- Ask recreation and health Professionals for more information about available facilities and scheduled events

This is a GOOD START !

GET IN THE FLOW

Following are some of the thoughts of Mihaly Csikszentmihalyi found in his NATIONAL BESTSELLER — " FLOW ...THE PSYCHOLOGY OF OPTIMAL EXPERIENCE – STEPS TOWARD ENHANCING THE QUALITY OF LIFE." I first heard of his book while taking a correspondence course called, "BORN TO LEARN." It is my hope that you will read "FLOW." It is my greater hope that you will EXPERIENCE FLOW *!*

In his introduction Mihaly writes, "Twenty-three hundred years ago Aristotle concluded that, more than anything else, men and women seek happiness. While happiness itself is sought for it's own sake, every other goal — health, beauty, money, or power — is valued only because we expect that it will make us happy."

After decades of research on people and what they thought and felt during their happiest experiences Mihaly and his colleagues developed the concept of "FLOW." The state in which people are so involved in an activity that nothing else seems to matter. The experience itself is so enjoyable that people will do it for the sheer sake of doing it.

He says it's about getting control of your life. *Being in an Optimal Experience*, FLOW, *gives us a sense of participation in determining the content of our life.* Whenever the goal is to improve the quality of life, the FLOW theory can point the way. Optimal experiences are something that we can make happen.

For each person there are thousands of opportunities, and challenges to improve ourselves. Mihaly's research found that these Optimal moments, our best moments, happiest moments, usually occur when a person's body or mind is stretched to it's limits in a voluntary effort to accomplish something difficult and worthwhile. What really satisfies people, makes them happy, is not getting slim or rich, but feeling good about their lives. It is not just a matter of losing a few extra pounds. It is a matter of losing the chance to have a life worth living.

Please *!* Do yourself a favor and read Mihaly's book, "FLOW."

IT IS TIME!

People know that being fat is not only life threatening, it is costly. *Being fat is deadly.* It takes the fun out of life. As a parent, if you are fat, you're putting your children's lives in jeopardy.

Being in fit shape gives you more years in your life, and more importantly, it gives you more life in your years.

It takes work to be in shape. It's not easy. You have to *"KNOW SWEAT." It is a lifetime trip, a lifestyle.*

Exercise is necessary and what you do and don't eat is important. Diets don't work.

The body is smart. If you starve it, it will shut down, lower your metabolism, and hold onto it's fat.

This book, hopefully, will be a *"MOTIVATOR"* to make it clear that *being fat is so frightening, so costly, so painful physically and mentally* that the reader is moved into moving.

The rewards are priceless!

Give this book to a fat friend. *Show them that you care.* Save them and their children and their children's children from lots of expense, from stress – physical and mental, from sickness and pain, and a premature death.

We would all warn a person that is unaware – that can't see that they are going down a dangerous path. We've all been, at some time in our lives, unable to see some of the hazards ahead of us.

*Denial is **not** a river in Egypt.*

Hopefully, this book's message will shine a light bright enough for your friend to see the frightening consequences of being overweight.

Those folks who's *"give a damn"* isn't broken, will thank you for the gift.

Repetition has been used in this book because it is one of the best teaching tools. *Repetition – that's what practice and rehearsal mean.* Do it again. Play it again. Read it again. ..until you get it down... *FIRST THE IDEA AND THEN YOUR WEIGHT!*

TELEVISION WORTH WATCHING

Television, a useful and powerful teaching tool is being used more and more to encourage kids to eat right, be active, get along with others and get enough rest.

"Sesame Street" is getting parents to watch TV programs with their children and to become better role models by having celebrities appear on their shows. Rosemarie Truglio, Vice President of Education and Research for Sesame Street Workshops, thinks that *the obesity epidemic in our country is a "health crisis."* Including *healthy messages* in their television shows is a good start to solving the problem.

The Disney Channel offers "JoJo's Circus." This show features an animated clown that encourages kids to follow her, not unlike the Pied Piper. The clown asks preschoolers to skip, walk and dance with her, while watching the show. Anne Wood, creator of "Teletubies," has come up with "Boobah," a show featuring "Fuzzballs" meant to motivate the young audience to giggle and imitate them as they exercise. Like Julie Andrews sang "A little bit of sugar helps the medicine go down." They get the kids to exercise without calling it exercise.

The Nickelodeon Network Pre-school programming, "Nick Jr." has imported a show from Iceland. That's right Iceland. It is called "LazyTown." It is the brain child of world-class athlete and entertainer, Magnus Scheving. It was a hit in Iceland for eight years before coming to Nickelodeon. Magnus plays "Sportacus," a superhero, that never stays still. He is a super-healthy hero that always has lots of sports equipment, fruits and vegetables on hand. As protector of good health to the citizens of LazyTown, Sportacus is as popular with children of all ages as Superman, Batman, Tarzan and Mickey Mouse wrapped into one super hero. In 1991, Magnus and his "team" of one hundred sixty-three members created "LazyTown" and all its characters in a book "GO*!* GO LAZYTOWN*!*" to promote physical activity in the "Golden Age of Grease" (obesity epidemic). You might say his imaginative show has been successful…the book led to two musicals, a branded bottled water, shoes, toothpaste, beach toys, coloring books, t-shirts and a twenty-four hour radio station*!* As a physical educator, Magnus and Sportacus are *my super hero.* Yes, singular – *super hero.* To quote Magnus, *"Sportacus is part of me. It's who I am. I still put on the costume every week."*

Dr. Thomas N. Robinson, Director of The Center for Healthy Weight, at Lucille Packard Hospital, Stanford University, says "There is little known on exactly what approaches will be the most effective." He helped produce an Institute of Medicine report that *encourages the media to help solve the problem of childhood obesity by promoting good nutrition, increasing physical activity and to reduce sedentary behavior.*

It is my belief that the television shows mentioned are excellent evidence that the television media is demonstrating very positive action toward solving our country's obesity epidemic.

GOOD THINGS HAPPENING

A half century ago a writer/producer/ actor/ director/author, Don McGuire wrote a book *"THE DAY TELEVISION DIED."* The main character in the book was a "genius" TV advertising agency's idea guy named Marvin Lazarus.

The ad agency Marvin worked for, handled the account for the food company that bought more TV ad time than all the other companies advertising their products on TV.

The food company paid for most of the advertising space on most of the shows on this certain network. Most of the TV audience was watching shows on the *other* network. *What to do? Call "Marvin the Genius" he will know*!

Marvin's solution...*GO DARK – shut the network down*! Instead of showing silly sit-coms and dumb reality shows and horrifying news broadcasts, he suggested putting public service messages on the screen. Messages like: *go outside and play, go to the library, go for a bike ride, take a walk, go to the gym*!

After much persuasion, the network took Marvin's advice. The result – the network won the "rating race."

Now, fifty-some years later, October 1, 2005, *Nickelodeon went dark for three hours*! Their viewers were told to *go play, get active. CONGRATULATIONS* to Nickelodeon President, Cyma Zarghami and her staff*!*

~ LYNN SMITH
THE LOS ANGELES TIMES

I wish my friend Don McGuire had lived to see his vision come true!

100 KNOW SWEAT!

President George Bush has been cited as "one of our most fit presidents in years." His personal fitness routine includes mountain biking. He has said he is proud of the fact that he is in such good shape he leaves behind some of his special services men trying to keep up with him on the mountain bike course.

By setting this example it is no surprise the *"President's Challenge"* fitness program has registered 235,454 individuals. Over 10,000 schools are using the *Fitness File.* The top activities include *walking, running and biking.* So far, *28,052 people have earned the "Presidential Active Lifestyle Award."*

Wisconsin Governor, Jim Doyle, has launched the *Second Annual Wisconsin Governor's Challenge.* It is a six-week initiative to get more Wisconsin *residents up and moving.* Over 25,000 people have taken part in these friendly competitions.

President's Council member, John Burke, a Wisconsin native, was instrumental in the initial success and continued support of Governor Doyle. This has led to the start of the *Winter Challenge.* For more details see www.wisconsinchallenge.org.

Snack-food companies are under attack about childhood obesity. To address their critics, many food companies have introduced more healthy, reduced-sugar products and are also promoting exercise.

- PepsiCo is building thirteen playgrounds around the country as part of their campaign to promote exercise.
- Coca-Cola has donated four million dollars to eight thousand-five hundred middle schools to purchase pedometers.
- General Mills is giving two million dollars a year to schools and community groups for nutrition and fitness programs.
- Kraft Foods, Inc. will stop advertising its less-nutritious products on TV and radio and in magazines aimed at children under the age of twelve. They have also pledged six million dollars, over the next two years, for educational programs.
- McDonald's has signed up thirty-one thousand elementary schools to participate in "Passport to Play" programs that teach kids games around the world.
- Kellogg Co. gave two hundred seventy-five thousand dollars to teach kids fitness and nutrition. Messages appear on their packaging and on their web site asking kids and parents to do two things everyday: 1.) walk a mile and 2.) to eat breakfast – a bowl of cereal.

THIS IS A GOOD START!

The President's Council on Physical Fitness and Sport

Melissa Johnson, Executive Director, will be leading the P.C.P.F.S. into it's second half of a century. That's right – the Council's Fiftieth Anniversary is upcoming.

Lynn Swann, NFL Hall of Fame member, has recently retired as Chairman after years of exemplary leadership.

The P.C.P.F.S. has accomplished much since my father, Dr. Ben Miller, was involved during the Kennedy and Eisenhower Administrations.

Ms. Johnson welcomes visitors to the President's Council website: www.fitness.gov. One of the many ways the Council produces and distributes publications through this website is to partner with corporations and organizations, and other government agencies to produce timely and interesting publications for Americans of all ages, backgrounds and abilities. They have partnered with the Kellogg Company (http://kellogs.com/us/) and the National Association for Sport and Physical Education (NASPE), (http://www.aahperd.org/naspe/) to produce *Kids in Action: Fitness for Children to Age Five* (http://.gov/Reading_Room/Kidsinactionbook.pdf). The publication is available in both English and Spanish.

Another partnership is with the Blue Cross Blue Shield Association (BCBSA) to produce *Walking Works,* a guide to start and maintain a regular walking program.

They also partnered with the American Association of Clinical Endocrinologists to produce *Rx: Take The President's Challenge,* an English and Spanish physical activity prescription for physicians to distribute to children and adults.

P.C.P.F.S. adopted *Exercise: A Guide From the National Institute on Aging* as the physical activity publication for older Americans. (http://www.niapublications.org/exercisebook/ExerciseGuideComplete.pdf).

Dr. Leonard Epstein and his colleagues at Stanford University studies show that putting children on a diet decreasing their TV viewing, playing computer games, and talking on the phone helped the children lose weight. *Another diet that works.* Epstein also advises *"PARENTS MUST TAKE BACK CONTROL OF THE TABLE."*

After losing over one hundred pounds and keeping it off, Governor Mike Huckabee has become a role model for all the folks fighting the *"Battle of the Bulge."* Arkansas has since launched the *HEALTHY ARKANSAS INTIATIVE.* The program is simple…encourage healthy behavior and provide motivation by coming up with a whole list of incentives for the people who make these choices.

"Stop treating snake bites and start killing snakes." is how Dr. Fay Boozman, Director of the Arkansas Department of Health describes the program. The Governor's hope and goal is that a *Healthy Arkansas will lead to a "Healthy America."*

Incentives such as offering state employees, who are willing to undergo a health risk assessment to help measure the incidence of *risky behaviors*, will be given a discount of up to twenty dollars a month from their State Employee Health Insurance.
Preventative Medicine at work*!*

"Colorado On The Move," a weight management program, has been started by Jim Hill and his colleagues at the University of Colorado Health Sciences Center. The goal…to increase by two thousand steps a day the average number of steps the average Coloradan takes. They got the idea from The National Weight Control Registry, the most successful single group to maintain weight loss after several years. The low-tech step counters being distributed around Colorado are underwritten by a consortium of government agencies, private foundations, educational institutes and businesses. More than six thousand people are enrolled in the pilot program.

~ *GREG CRITSER*
"FAT LAND"

KNOW SWEAT! 103

Scientific research continues on the study of obesity, its causes, effects and cures. Researchers in the U.S. and Britain recently discovered that *fat cells speed up the aging process.* This study showed the more a person weighs, the older the cells appear on a molecular level.

Tim Spencer of the St. Thomas Hospital in London, who led the study that was published on-line by the Lancet Medical Journal says, "We have known obesity increases your risk of many diseases, and of dying early. What's novel here is that it seems that *fat actually accelerates the aging process."* *The study showed that obesity adds nine years of age to a person's body.*
~ THE WASHINGTON POST

Ronald McDonald will be wearing a sweat suit in some of McDonalds' television commercials and print ads. McDonalds offers a very healthy selection of foods in their salads. Their *"GO ACTIVE"* Happy Meal for adults includes a salad, water and a pedometer.

Sesame Street, one of the best "home classrooms" for over thirty-five years is creating several new puppets to their cast representing healthy vegetables. The Cookie Monster will not go on a diet but will use humor to get healthy messages to the kids. He will be eating healthy foods such as broccoli.

Dr. Julie Gerberding, Chief of the Center for Disease Control and Prevention (CDC), faults a recent report on people being overweight that concluded that mildly overweight people had a 20% lower risk of dying than those who weigh less.

The report also stated that obesity accounted for 25,814 deaths a year, hugely lower than the 365,000 deaths estimated just a few months earlier. The study made by the CDC, our nation's health agency, has caused confusion with it's controversial conclusions.

"It is not okay to be overweight. People need to be fit, they need to have a healthy diet, they need to exercise," says Dr. Gerberding. *"I am very sorry for the confusion that these scientific discussions have caused."* Good for Dr. Gerberding!
~ ALBUQUERQUE JOURNAL
ASSOCIATED PRESS

Dan Latham is a Physical Education Teacher. You know, one of those PE majors that the other students laugh at and cast aspersions (make fun of...) "Dumb PE major, jock!" (like that!)

Many kids became overweight by eating too much junk food, drinking too many soft drinks, watching too much television, playing too many video games and not exercising enough.

No one on the faculty could come up with an idea to get the kids to exercise. Dan realized that there was an unused two thousand square foot building on campus that students now wait to get in to. Why? To Exercise!

Currently, there are fifty-five pieces of equipment that range from Stair Steppers to Stationary Bicycles, from Videogame Bikes to Interactive Dance Games. They are wired to big screen video games. The power to make the characters respond, to be able to play the games, is generated by the kids pedaling the bikes. When they stop pedaling, the game becomes inactive.

These same kids that the coaches , their teachers, even their friends couldn't get to walk around the block, much less pedal a bike of any kind, are sweating and having fun doing it.

Dan has named his fitness center "CYBEROBICS." The center can accommodate up to fifty-five students.

As a result of this programming, during the 2001–2002 school year, West Middle School, in Downey, California, registered its biggest gains on the State Fitness Gram testing assessment. Students at West Middle School were NUMBER ONE in the category of aerobic capacity.

Not a bad idea coming from a "dumb" PE major. We all should be fortunate enough to be taught by as creative a teacher as Dan Latham.

Ninety-seven is the percentage of U.S. adults who don't follow all four rules for healthy living: *eating right, keeping a healthy weight, exercising, and not smoking.*
                                        ~ AARP The Magazine

"WE CAN DO BETTER THAN THIS*!*"

~ DR. SEUSS

BEEN THERE — DONE FAT!

"One might say of me that in this book I have only made up a bunch of other people's flowers, added a few of my own and provided the string that ties them together."

~ *MONTAIGNE*

BIBLIOGRAPHY

Allen, Steve	"DUMBTH" The Lost Art of Thinking
Anderson, Bob	"STRETCHING"
Benson, Herbert M.D.	"THE RELAXATION RESPONSE" "BEYOND THE RELAXATION RESPONSE" "YOUR MAXIMUM MIND"
Bierce, Ambrose	"THE DEVIL'S DICTIONARY"
Cantu, Robert C. M.D.	"REGAINING HEALTH AND FITNESS"
Cooper, Kenneth H. M.D.	"THE AEROBIC'S WAY"
Cosby, Bill	"I AM WHAT I ATE...AND I'M FREGHTENED"
Cousins, Norman	"ANATOMY of an ILLNESS as PERCEIVED by the PATIENT"
Covey, Stephen R.	"THE 7 HABITS OF HIGHLY EFFECTIVE PEOPLE POWERFUL LESSONS IN PERSONAL CHANGE"
Critser, Greg	"FATLAND" "HOW AMERICANS BECAME THE FATTEST PEOPLE IN THE WORLD"
Csikszentmihalyi, Mihaly	"FLOW" The Psychology of Optimal Experience
Davis, Adele	"LET'S EAT RIGHT TO KEEP FIT"
Diamond, Harvey and Marilynn	"FIT FOR LIFE"
Dyer. Dr. Wayne W.	"PULLING YOUR OWN STRINGS"
Emmerton, Bill	"THE OFFICIAL BOOK OF RUNNING"
Emerson, Ralph Waldo	"SELF-RELIANCE"
Epstein, Leonard M.D.	"STOPLIGHT DIET"
Garfield, Charles A.	"PEAK PERFORMERS"
Gellis, Dr. Marilyn	"MENTAL FLOSS"

Giono, Jean	"THE MAN WHO PLANTED TREES"
Greene, Bob	"GET WITH THE PROGRAM"
Grudin, Robert	"TIME AND THE ART OF LIVING"
Hill, Dr. James O.	"THE STEP DIET BOOK"
Huckabee, Gov. Michael	"QUIT DIGGING YOUR GRAVE WITH A KNIFE AND FORK" "A 12-STOP PROGRAM TO END BAD HABITS AND BEGIN A HEALTHY LIFESTYLE"
Johnson, Spencer M.D.	"WHO MOVED MY CHEESE?"
Keillor, Garrison	"THE BOOK OF GUYS"
Kuntzleman, Charles T.	"RATING THE EXERCISE"
Langer, Ellen J.	"MIND-FULNESS"
Lorig, Kate R.N. Dr.P.H. and Fries, James F. M.D.	"THE ARTHRITIS HELP BOOK"
McGraw, Dr. Phil	"SELF MATTERS" AND "THE ULTIMATE WEIGHT SOLUTION, THE 7 KEYS TO WEIGHT LOSS FREEDOM"
McGuire, Don	"THE DAY TELEVISION DIED"
McWilliams, Peter and John-Roger	"FOCUS ON THE POSITIVE" "LIFE 101" "YOU CAN'T AFFORD THE LUXURY OF A NEGATIVE THOUGHT"
Miller, Ben Ph.D., Bookwalter, Karl Ed.D, and Schlafer, George M.S.	"PHYSICAL FITNESS FOR BOYS"
National Institute on Aging	"EXERCISE: A GUIDE FROM THE NATIONAL INSTITUTE ON AGING"

Nelson, Miriam, Ph.D.   "STRONG WOMEN AND MEN BEAT ARTHRITIS"

Peter, Dr. Laurence J.   "THE PETER PLAN"

Pollack – Wilmore – Fox   "EXERCISE IN HEALTH AND DISEASE"

Prudden, Suzy   "I CAN EXERCISE ANYWHERE BOOK"

Rowe, John W. M.D.   "SUCCESSFUL AGING"
and Kahn, Robert L. Ph.D.

Schwarzenegger, Gov. Arnold   "ARNOLD'S BODY SHAPING FOR WOMEN"
                                  "ARNOLD: THE EDUCATION OF A BODYBUILDER"

Selye, Hans   "STREE WITHOUT DISTRESS"
                                  "THE STRESS OF LIFE"

Seuss, Dr.   "OH! THE PLACES YOU'LL GO!"
(AKA Theodor Geisel)

Time Life Books   FITNESS, HEALTH and NUTRITION SERIES
                                  "THE BODY IN MOTION"
                                  "SUPER FIRM"
                                  "WALKING AND RUNNING"
                                  "THE FIT BODY"
                                  "WELLNESS"
                                  "CROSS TRAINING"
                                  "RESTORING THE BODY"
                                  "QUICK WORKOUTS"
                                  "SOFT WORKOUTS"
                                  "STAYING FLEXIBLE"

U.S. Air Force Publication   "THE U.S. BOOK of FAMILY PHYSICAL FITNESS"

Weil, Andrew M.D.   "SPONTANEOUS HEALING"

Willett, Walter M.D.   "EAT, DRINK and BE HEALTHY"

ABOUT THE ILLUSTRATOR

Character artist and product designer, Mike Royer, born and raised in Oregon, was lured to Southern California in the Spring of 1965, to pursue a career in comic art. He spent the next fourteen years contributing to the world of comic books, comic strips, and TV animation. He worked on a variety of subjects including MAGNUS, ROBOT FIGHTER, TARZAN and SPACE GHOST, in comic books, coloring books, and puzzles for Western Publishing Company, publishers of Gold Key Comics and did layouts on one-third of the first network animated SPIDERMAN Saturday Morning Series.

While continuing to draw and sometimes write such properties as SPEED BUGGY, H-B TV ADVENTURE HEROES, TARZAN and MAGNUS, Mike contributed to James Warren's CREEPY, EERIE and VAMPERELLA magazines. He began drawing and writing the comic panel CRUSIN' record album covers – over two dozen at last count.

During the 1970s Mike was best known to comic art fans as the letterer/inker for the legendary JACK KIRBY at DC and Marvel comic books. In the late 1970s Mike inked both the TARZAN and STAR WARS comic strips.

In the Spring of 1979, Mike joined the staff at the WALT DISNEY COMPANY, in the Creative Department for their Consumer Product/Licensing Division, addressing the areas of book publishing, comic books and strips, and the various types of theme park and licensed merchandise as a Character Artist/Product Designer as "idea man," concept and final line artist. In 1993 Mike created the *"new look"* that launched massive WINNIE THE POOH licensing program. Featured in a forty-three minute video "HOW TO DRAW POOH", sent to more than forty licensees, Mike takes great pride in the fact that Pooh soon was outselling Mickey Mouse (previously the character most drawn by Mike) worldwide. Mike spent the next seven years as *the main Pooh man* for Disney Stores Creative Division, finding time to do 3-D products for Disney and Warner Brothers stores.

In 2001, Mike and his lovely wife and idea collaborator, Laurie, moved back to Southern Oregon. Since embracing the setting and lifestyle of the beautiful Rogue Valley, Mike has found time to ink some comic books for Marvel and DC and continues to create Disney Limited Edition Collector Pin Sets graphics, such as the Winter Olympic Games 2006, featuring Mickey Mouse and the gang. The past six years Mike has functioned as an art service on a wide variety of projects including orthographic turns and "floor plans" for computer game animator, DIGIMON products, READER RABBIT and BABY FAITH books, RESCUE HEROES toy packaging, and more.

In 2005, Mike met Denny Miller. Since starting the drawings for this book Mike has lost thirteen pounds and strongly believes that Denny's passion, attention to detail and accuracy, and his commitment to integrity *and a healthy lifestyle* will keep Mike at the drawing board for a long time. More on Mike's works at www.michaelroyer.com.

ABOUT THE AUTHOR

Denny Miller has a degree in Physical Education from the University of California, at Los Angeles. He has been a personal trainer – "Me Tarzan, You Train" is the title of his brochure. He was a Personal Trainer in his own gym, "The House of Reps," and also taught fitness and relaxation at California Spas. He taught relaxation at the University of California, at Humboldt; and at Fort Lewis College, in Durango, Colorado; for The American Arthritis Foundation at the Annenberg Center, Eisenhower Medical Center, Rancho Mirage, California; and for the U.S. Navy at Point Mugu, California.

In 1983, Miller wrote, produced, directed and was the on-camera instructor for an award winning relaxation video "HOMESTRETCH." This video received First Place honors from the Institute of Creative Research and The Sport Art Academy (NASPE) at the 1991 AAHPERD National Convention, in San Francisco.

"HOMESTRETCH" uses the relaxation response technique developed by Dr. Herbert Benson, Harvard Medical School. Slow stretching, soft lights, soft music and mind and breath control with practice, will bring on the relaxation response.

While at UCLA, Denny had the privilege of playing Basketball for the legendary Coach John Wooden.

During his forty-nine year acting career he starred or co-starred with many of Hollywood's biggest names, in 235 television episodes, including almost four years on Wagon Train, as the Scout, Duke Shannon. He has acted in twenty films including the starring role in "TARZAN THE APE MAN," in 1959, for MGM. He has been featured in more than 200 television and radio commercials. He was cast as the spokesman for Gorton's as the Gorton's Fisherman for fourteen years, until he was seventy years old.

Denny founded a publishing company, To Health With You Publishers in 2004, and released his first book, an autobiography, "DIDN'T YOU USED TO BE...WHAT'S HIS NAME?" the same year.

Denny Miller's profession is Health Educator and his career is Acting.